FAMOUS BRIDGES

Contents

Annette Smith

Bridge Designs

Bridges are designed to transport people, animals, vehicles and freight from one location to another. A bridge might cross small areas of water or land, cavernous valleys or hazardous seas.

People from the earliest **civilisations** made simple forms of bridges. They used natural materials, such as the trunks of fallen trees, which were dragged into position across streams, narrow rivers and small valleys.

Think and Talk About ...

Engineers today commonly use early bridge designs: beam (a simple bridge often made from a log or wood plank), arch (a curved bridge usually made from stone or brick) and suspension (a bridge that is hung below cables).

Stone bridges were built with large rocks to make roads over uneven ground for animals and carts.

Forest vines and creepers were twisted together to form strong ropes. The ropes were suspended high above deep **ravines**, allowing people to cross quickly and safely.

Arch bridges were originally made from large stones.

Thousands of years ago, Greeks and Romans built huge **aqueducts** from stone, to transport water needed for drinking, bathing and irrigating crops. Several of these massive constructions can still be seen in parts of Europe and the United Kingdom.

Many bridges today, and from years gone by, have become famous for their design, as well as the skills and materials used for their construction.

The Stari Most ("old bridge") in Bosnia was built in the 17th century.

The Pontcysyllte Aqueduct in Wales is Britain's longest aqueduct.

The Akashi Kaikyo suspension bridge in Japan is famous for its design.

Pont du Gard Nimes, France

One of the most famous bridges in the world is the Pont du Gard, in southern France. It is a three-level stone aqueduct, built by the Romans more than 2000 years ago. It was constructed to transport water over the Gardon River Valley to the city of Nemausus. Today, the city is known as Nimes.

The Pont du Gard has 52 arches.

The Pont du Gard is an arch bridge. The design has many arches because the bridge is so wide. The arches spread the weight of the long beams above, making the bridge very strong.

The Pont du Gard was constructed with blocks of **limestone** rock. The rock came from a nearby quarry, on the banks of the river. These huge blocks were held together with **clamps** made of iron. The bridge was believed to have taken 1000 workers three years to build.

Today, the Pont du Gard is not used as an aqueduct or for transporting vehicles. It has become a major tourist attraction.

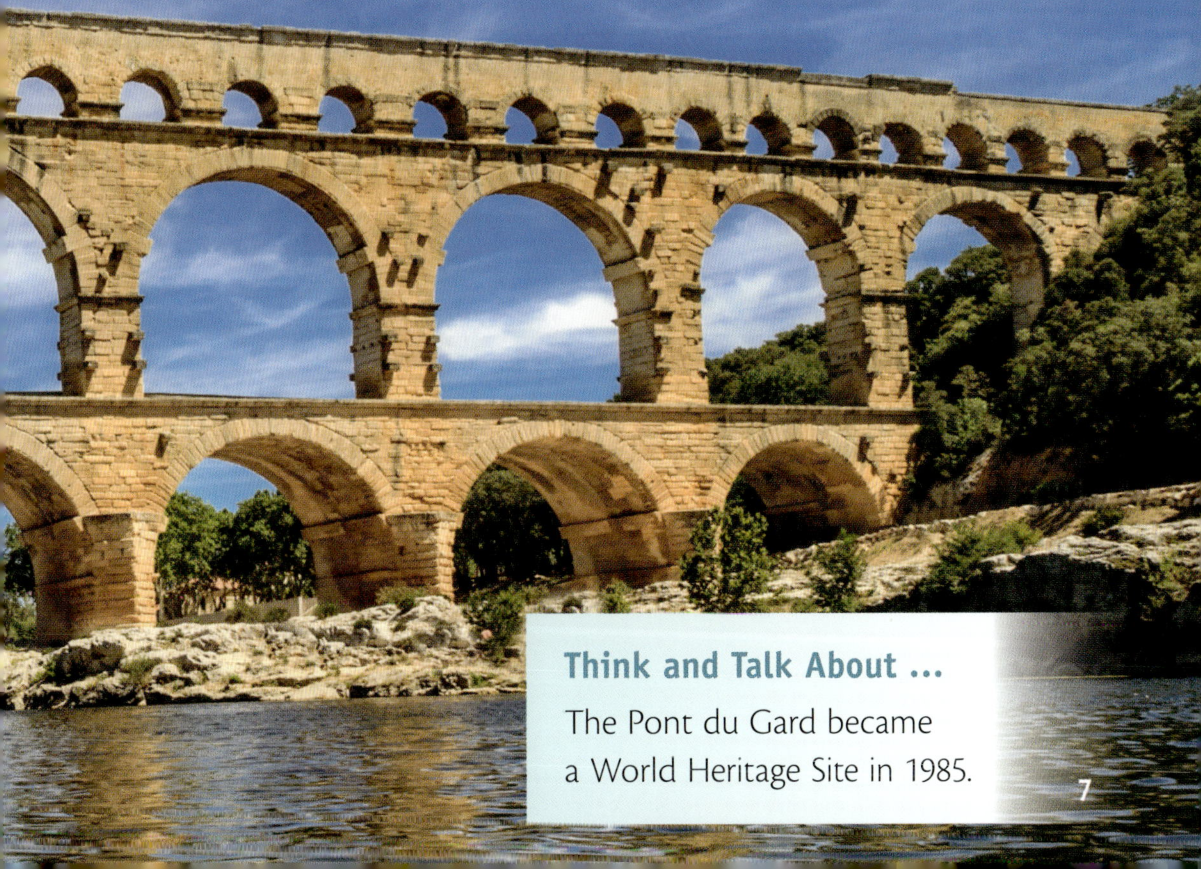

Think and Talk About ...

The Pont du Gard became a World Heritage Site in 1985.

Chapel Bridge Lucerne, Switzerland

Chapel Bridge is the oldest covered, wooden footbridge in Europe. It crosses the Reuss River in the northern city of Lucerne, Switzerland. The bridge was built in 1333 as a **fortification**. It linked the old town on the right bank of the river to the new town on the left bank. This helped to protect the city from raiders coming from the south, near Lake Lucerne.

The water tower (left, next to Chapel Bridge) still stands today.

Standing in the river, alongside the bridge, is a 40-metre tall *Wasserturm*, or water tower. It is an octagonal-shaped building made from brick. The tower does not contain water. It was built several years before the bridge and was originally used as a prison and a **watchtower**.

Chapel Bridge is unique because it has a large collection of paintings dating as far back as the 17th century. The paintings represent events in the history of Lucerne. The artist painted on wooden **panels** made from spruce, maple and linden trees. The triangular-shaped panels were set all along the interior pitched roof of the bridge.

In 1993, the bridge was almost completely destroyed by fire. As a result, most of the paintings could not be saved. Only 30 of the 147 that existed before the fire could be restored.

Chapel Bridge was rebuilt in 1994. Each year, thousands of people visit Lucerne to photograph the bridge and water tower, which have become a symbol of the city.

Tower Bridge London, England

The River Thames flows through the centre of the city of London, England. During the late 19th century, the city became densely populated, so a new bridge was needed to allow people to move quickly and easily across the river.

Tall ships can pass below Tower Bridge.

The city **architect**, Sir Horace Jones, designed Tower Bridge, which is now one of London's major attractions. The bridge took eight years to build and was officially opened in June 1894.

Tower Bridge is made up of two huge towers built on enormous concrete **piers**, which are sunk deep into the riverbed.

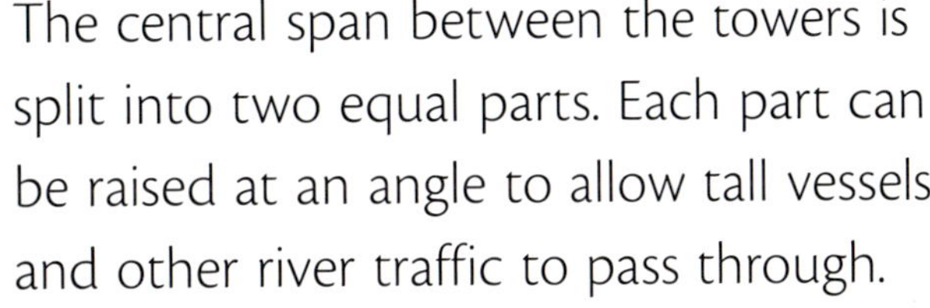

The central span between the towers is split into two equal parts. Each part can be raised at an angle to allow tall vessels and other river traffic to pass through.

In recent years, one of the high-level walkways between the towers has been transformed with new lighting and a glass floor. The glass floor provides visitors with a breathtaking view of the river and city below.

Today, more than 40 000 pedestrians, cyclists and motorists use Tower Bridge every day.

Think and Talk About ...

Tower Bridge in London is a combined bascule (moveable bridge) and suspension bridge.

Chengyang Wind and Rain Bridge
Sanjiang County, China

The famous Chengyang Wind and Rain Bridge is located in Sanjiang County, China. It crosses the Linx River and was built in 1916.

Chengyang Bridge is a covered wooden structure supported by five large stone piers. The corridor along the length of the bridge has benches and rooms for people to meet and rest. When the weather is poor, people are protected from the wind and rain.

Above each pier is an ornately designed wooden tower with several verandas on different levels. The verandas and towers have wing-like horns, which look like birds flapping their wings. The entire roof area is covered with tiles.

Chengyang Bridge was built without the use of nails or metal **rivets**. Instead, workers skilfully cut and wedged pieces of timber together. Even today, more than a century later, this beautiful bridge still remains as strong as when it was first constructed.

Sydney Harbour Bridge Sydney, Australia

Sydney Harbour Bridge is one of Australia's most well-known landmarks. It spans the Sydney Harbour and is the world's tallest steel arch bridge. The bridge connects the business part of the city to the North Shore, where many people live.

Construction of the bridge began in 1924, and it took eight years to build. During the time the bridge was being built, an underground railway system was also being planned for the city. The bridge was designed to have six lanes of road traffic, with a railway line on each side and a walkway for pedestrians. Both sets of rail line were linked to the underground railway station.

Giant "creeper" cranes were used for the construction of the huge steel arch. The cranes were positioned on each side of the harbour and used to hoist workers and materials onto the two sections of the arch. Once the steelwork for the arch was completed, work on the roadway began.

On Sydney Harbour Bridge today, there are eight lanes for vehicular traffic and two rail tracks. Pedestrians have a walkway on one side of the bridge and cyclists have a bike lane on the other side.

Sydney Harbour Bridge attracts thousands of tourists from around the world. Visitors come to not only view the remarkable construction of the bridge, but also to experience many events that take place on it, such as the bridge climb and the fireworks on New Year's Eve.

The Sydney Harbour Bridge is one of Australia's biggest tourist attractions.

Think and Talk About ...

The top of the steel arch on the Sydney Harbour Bridge rises and falls up to 180 centimetres due to changes in the air temperature.

Golden Gate Bridge San Francisco, USA

The Golden Gate Bridge in the USA is sometimes described as a modern wonder. It is one of the world's longest bridges. It spans a vast area of sea from the city of San Francisco to Marin County, near the Pacific Ocean.

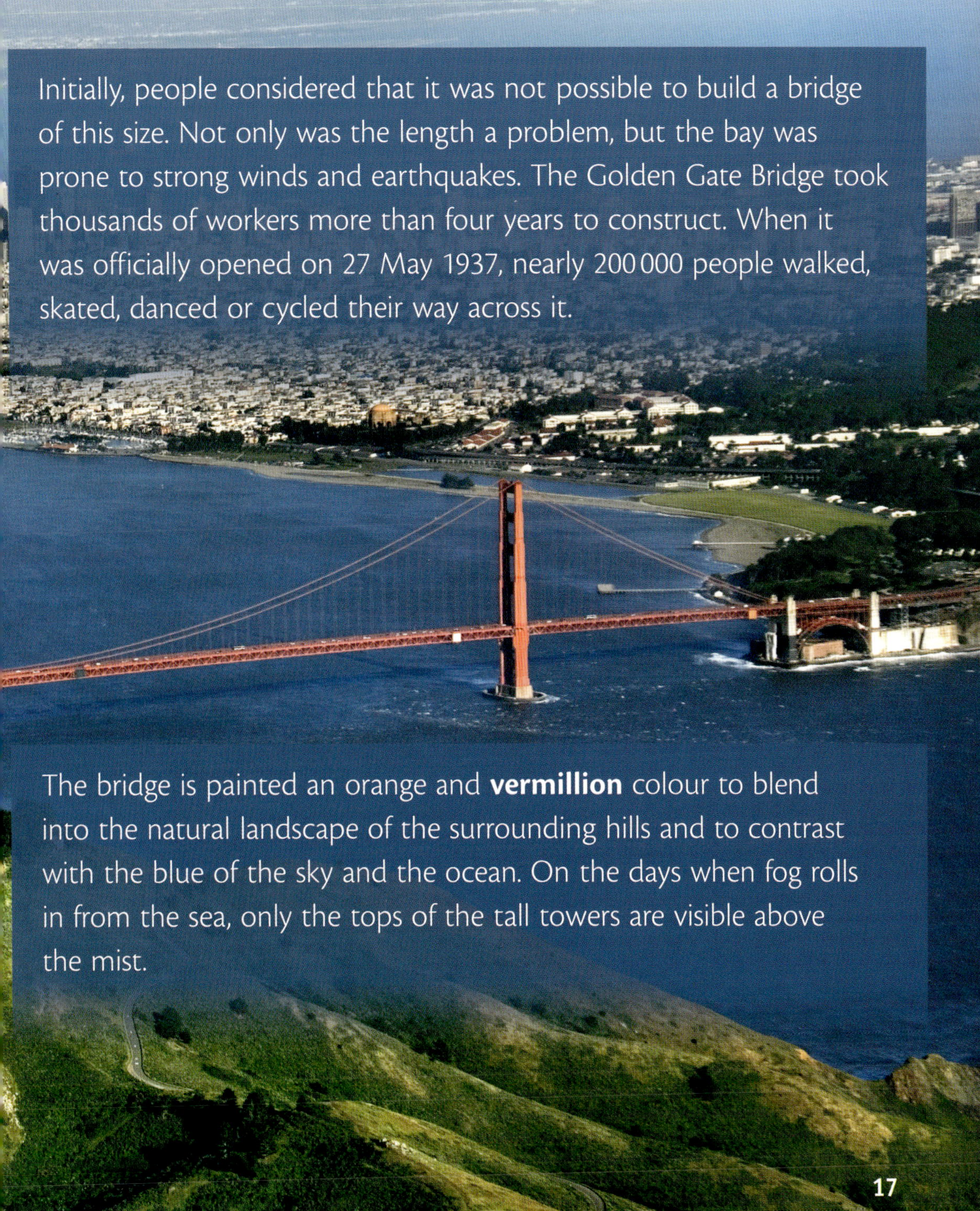

Initially, people considered that it was not possible to build a bridge of this size. Not only was the length a problem, but the bay was prone to strong winds and earthquakes. The Golden Gate Bridge took thousands of workers more than four years to construct. When it was officially opened on 27 May 1937, nearly 200 000 people walked, skated, danced or cycled their way across it.

The bridge is painted an orange and **vermillion** colour to blend into the natural landscape of the surrounding hills and to contrast with the blue of the sky and the ocean. On the days when fog rolls in from the sea, only the tops of the tall towers are visible above the mist.

Oresund Bridge Denmark and Sweden

Oresund Bridge crosses a vast expanse of sea between the countries of Denmark and Sweden. The bridge extends nearly eight kilometres from the coast of Sweden to an artificial island, Peberholm. From there, it links with a 4-kilometre underwater tunnel to the coast of Denmark.

Oresund Bridge is the longest combined road and rail bridge in Europe. The construction of the bridge began in 1995 and was completed four years later. It is a huge structure with two railway tracks beneath the four-lane traffic roadway. Two pairs of towers in the middle of the bridge are supported with cables. The towers are high enough for ships to pass safely beneath the main span of the bridge.

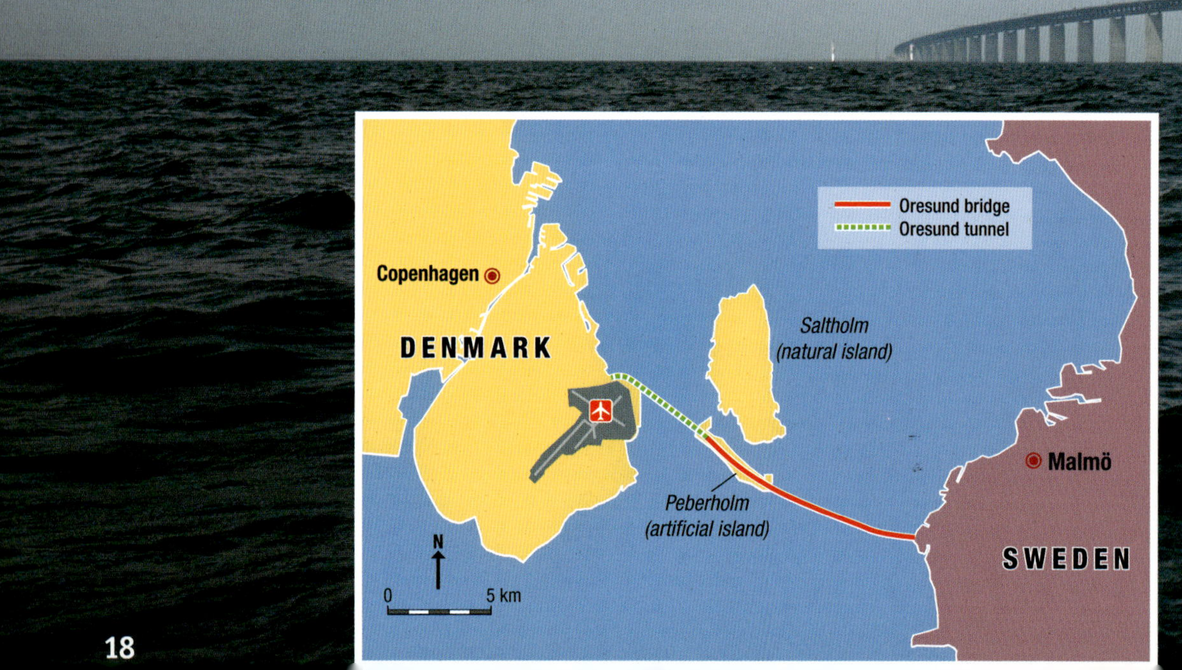

Oresund Bridge joins the underwater tunnel on Peberholm Island. The tunnel is designed as a tube tunnel. Two tubes have rail tracks and two carry road traffic. A fifth tube is smaller and is used only for emergencies.

People now have the opportunity to travel frequently and easily between the major cities of Copenhagen in Denmark and Malmö in Sweden.

The Oresund Bridge connects Denmark and Sweden.

Think and Talk About ...

The flight path for aeroplanes flying to Copenhagen airport is directly over the underwater tunnel.

Millau Viaduct Millau, France

The Millau **Viaduct** crosses the Tarn Valley in France. It was built to connect Paris in the north of France with the city of Perpignan in the south. When it was completed in 2004, it was the highest road bridge in the world.

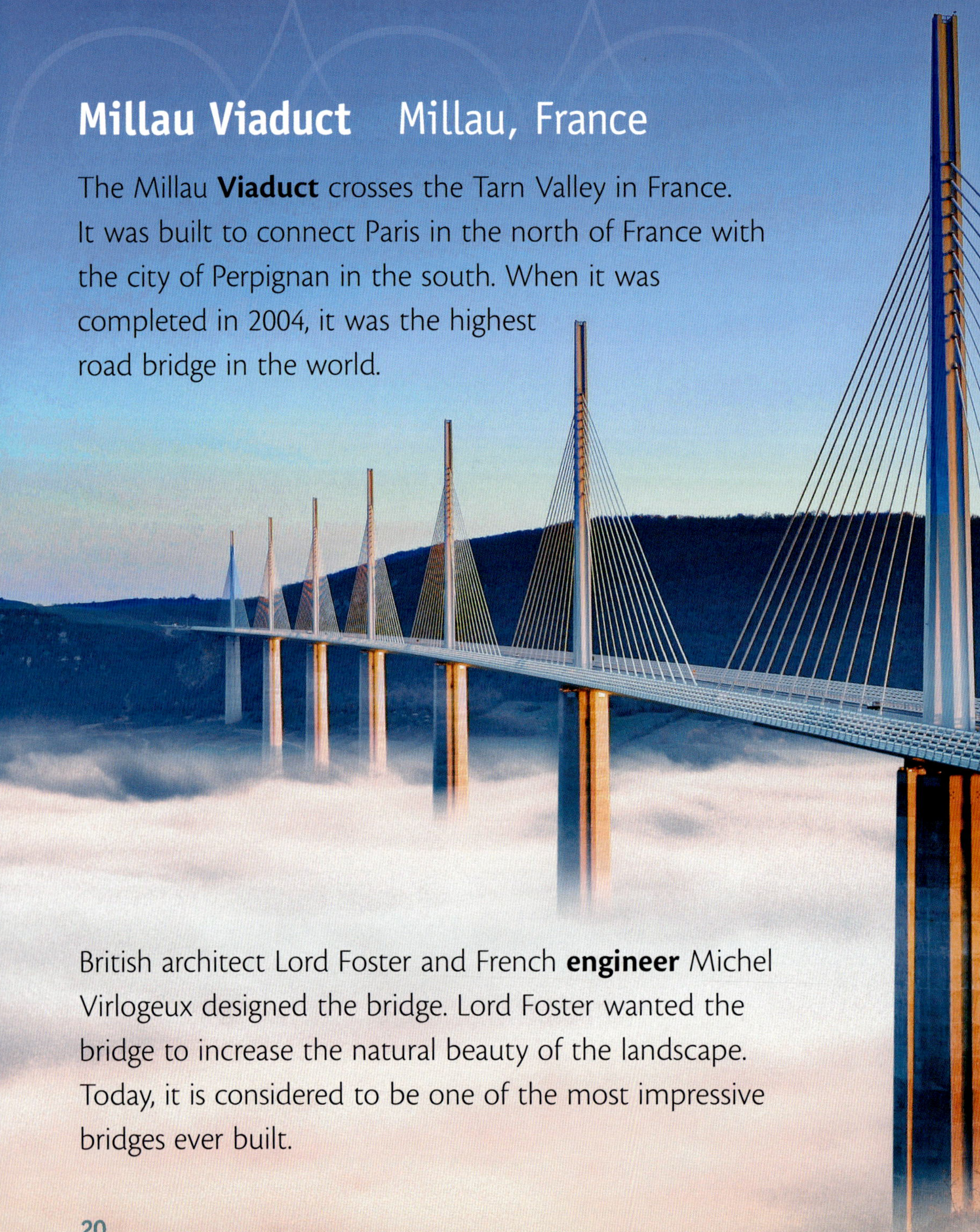

British architect Lord Foster and French **engineer** Michel Virlogeux designed the bridge. Lord Foster wanted the bridge to increase the natural beauty of the landscape. Today, it is considered to be one of the most impressive bridges ever built.

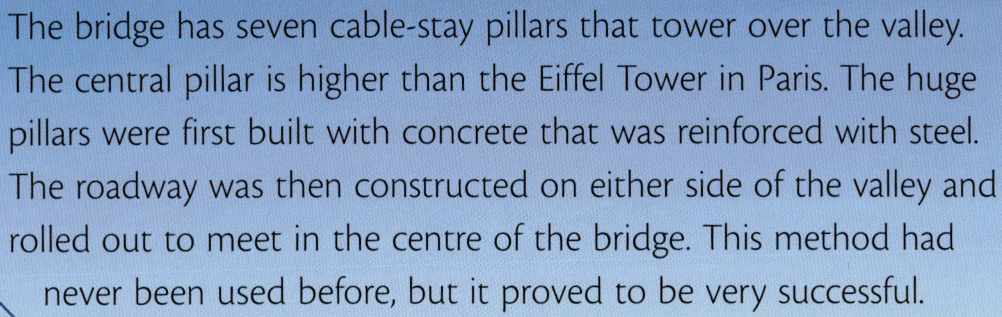

The bridge has seven cable-stay pillars that tower over the valley. The central pillar is higher than the Eiffel Tower in Paris. The huge pillars were first built with concrete that was reinforced with steel. The roadway was then constructed on either side of the valley and rolled out to meet in the centre of the bridge. This method had never been used before, but it proved to be very successful.

The Millau Viaduct is often photographed early in the morning, when the pillars appear to look like seven ships sailing on a sea of mist.

Think and Talk About ...

Vehicles are charged tolls when crossing the Millau Viaduct.

Sheikh Zayed Bridge
Abu Dhabi, United Arab Emirates

The Sheikh Zayed Bridge is an outstanding **feat** of modern architecture. The bridge is 842 metres long and spans the Maqta Channel, linking the island of Abu Dhabi to the United Arab Emirates' mainland. It was built to improve the traffic flow across this very busy area of land and water. The bridge was officially opened in November 2010.

The bridge was designed to withstand wind gusts of 160 kilometres an hour, strong sea currents and heavy traffic. Tonnes of concrete and steel were used in its construction. The bridge has three main arches made of huge concrete blocks that are strengthened with steel wires. The central arch is 63 metres high. The arches resemble giant waves curving up and down in sand dunes.

The magnificent lighting system on the bridge highlights the beauty of the design.

Changing Designs

Bridges are constructions that have a unique purpose. They allow people to connect to and explore new places on Earth. The design of a bridge depends entirely on its location and the purpose for which it will be used.

Today, some bridges are famous for the beauty and features of the design. Others are admired for the quality and **longevity** of the construction.

Bridge designs are constantly changing as building materials and expertise continue to develop.

the Charles Bridge in Prague, Czech Republic

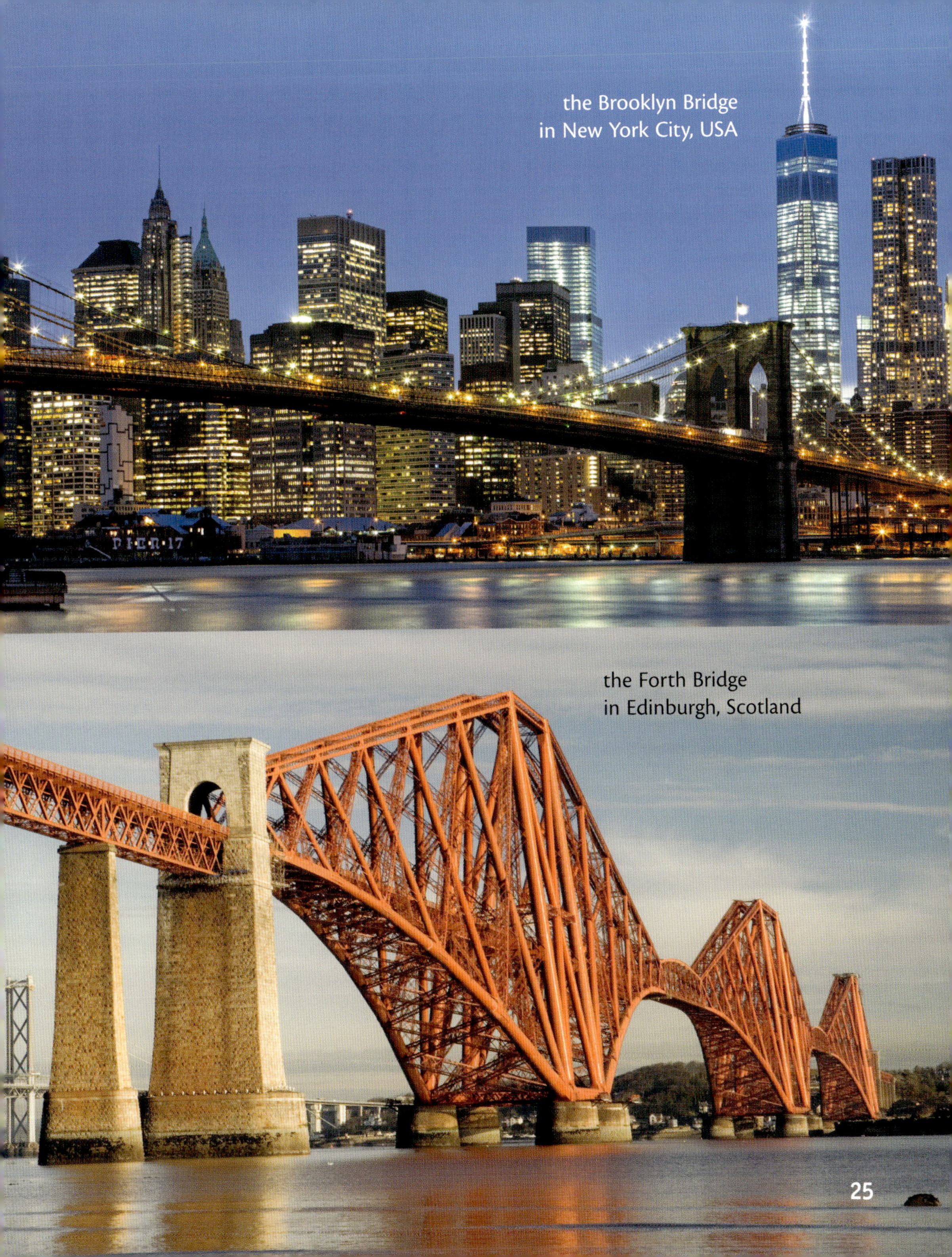

the Brooklyn Bridge
in New York City, USA

the Forth Bridge
in Edinburgh, Scotland

Ponte Vecchio: A Magical Place

When travelling in Italy, the historic bridge, Ponte Vecchio, is a magical place to visit. This **medieval** bridge crosses the River Arno in the city of Florence. It is believed that the original bridge, built during Roman times, was destroyed by raging floodwaters. Another bridge was constructed, but it, too, was destroyed when the river burst its banks.

Think and Talk About ...

The Grand Duke of Tuscany had a covered corridor built above the bridge in 1565, so he could walk safely from his palace to his office in the city.

The existing bridge was rebuilt in 1345 with three arched spans made of stone. The covered wooden walkway has small houses protruding from the sides, which was a common building method in large European cities during this period. Initially, the houses were used as workshops by butchers and **tanners**.

Ponte Vecchio, built in 1345, still stands today.

Tourists and locals can buy art, souvenirs and many other items on Ponte Vecchio.

Today, the houses on Ponte Vecchio are occupied as shops by art dealers, jewellers and souvenir sellers. Tourists enjoy the vibrant **atmosphere** that radiates from the merchants and artists attempting to sell their goods to anyone who comes near. Tables are set up outside the shops and along the length of the covered walkway of the bridge. Food stalls with delicious slices of watermelon and cones of gelato are a welcome relief in the heat of the mid-summer sun.

On a clear day, and even in the pouring rain, Ponte Vecchio is a beautiful setting for photography. The evening lights from the city create perfect reflections on the archways of the bridge and on the ripples of the water. Visitors to Florence find Ponte Vecchio a breathtaking place to experience.

Glossary

aqueducts (*noun*)	bridges that carry water over valleys
architect (*noun*)	a person who designs bridges and buildings
atmosphere (*noun*)	the mood of a place
civilisations (*noun*)	societies that have their own culture
clamps (*noun*)	objects that are used to hold two things together
engineer (*noun*)	a person who builds bridges, roads and machines
feat (*noun*)	something that is a great achievement
fortification (*noun*)	a wall that is built to protect a place from attack
limestone (*noun*)	a white rock that is used for building
longevity (*noun*)	how long something lasts
medieval (*adjective*)	something that was made in the period between 1100 CE and 1500 CE
panels (*noun*)	flat pieces of wood
piers (*noun*)	structures that stretch out into the water
ravines (*noun*)	deep, narrow valleys
rivets (*noun*)	pins that are used to fasten sheets of metal together
tanners (*noun*)	people who make leather
vermillion (*noun*)	a colour similar to red
viaduct (*noun*)	a bridge that carries a road over a valley
watchtower (*noun*)	a high tower that someone stands on and looks out for danger

Index

Famous Bridges

Text: Annette Smith
Series consultant: Annette Smith
Publishing editor: Simone Calderwood
Editor: Ted Carisbrooke
Project editor: Annabel Smith
Designer: Karen Mayo
Series designers: James Lowe and Karen Mayo
Photo researcher: Wendy Duncan
Production controller: Erin Dowling
Reprint: Siew Han Ong

PM Guided Reading
Emerald Level 25

Moving On
Famous Bridges
Hector's Electro-Pet Shop
The Arctic Circle
The Swimming Pool Project
Making Art with Light
Alex the Super Soccer Striker
The Great Barrier Reef
Popcorn the Wonder Pony
A Community Cares and Shares

ISBN 978 0 17 036890 2

Cengage Learning Australia
Level 5 , 80 Dorcas Street
Southbank VIC 3006
Phone: 1300 790 853
Email: aust.nelsonprimary@cengage.com

For learning solutions, visit cengage.com.au

Acknowledgements
We would like to thank the following for permission to reproduce copyright material:

Cover, p. 30: iStockphoto/Xavier Arnau; p. 1, p. 5 (bottom): Alamy/Motion/Horizon Images; pp. 2–3: Alamy/Peter Eastland; p. 4: iStockphoto/santirf; p. 5 (top): iStockphoto/Alasdair Thomson; pp. 6–7: iStockphoto/Bertl123; pp. 8–9: Shutterstock.com/Olgysha; pp. 10–11: Alamy/G Dobner; pp. 12–13: iStockphoto/anthonymadison; pp. 14–15: iStockphoto/kensorrie; pp. 16–17: amana images/Ocean/68; p. 18 (inset): Guy Holt © Cengage Learning Australia; pp. 18–19: iStockphoto/Niclasbo; pp. 20–21: amana images/Jean-Pierre Lescourret; pp. 22–23: Shutterstock.com/Rus S; p. 24: iStockphoto/banarfilardhi; p. 25 (top): iStockphoto/Maciej Noskowski, (bottom): iStockphoto/northlightimages; pp. 26–27: iStockphoto/MoreISO; pp. 28–29: Shutterstock.com/Brendan Howard.

Every effort has been made to trace and acknowledge copyright. However, if any infringement has occurred, the publishers tender their apologies and invite the copyright holders to contact them.

Printed in China by 1010 Printing International Ltd
14 25

This product is made from materials that are compliant with the EU Deforestation Regulation

People have been making bridges for thousands of years. The oldest bridges were made of tree trunks and rocks. Today, there are many bridges that are well known all over the world. Some are famous for their design, while others are famous for their engineering.

NELSON
A Cengage Company

Text Types:
Information Report (Informative)
Response (Informative)

ISBN: 978-0170368902
9 780170 368902